What Do We Know About the Chupacabra?

by Pam Pollack and Meg Belviso

illustrated by Andrew Thomson

Penguin Workshop

Para Jocia Alcala y Ariel Medina—PP

For Jena and Sean—MB

For Cerys and Rhia—AT

PENGUIN WORKSHOP
An imprint of Penguin Random House LLC, New York

First published in the United States of America by Penguin Workshop,
an imprint of Penguin Random House LLC, New York, 2023

Copyright © 2023 by Penguin Random House LLC

Penguin supports copyright. Copyright fuels creativity, encourages diverse voices, promotes free speech, and creates a vibrant culture. Thank you for buying an authorized edition of this book and for complying with copyright laws by not reproducing, scanning, or distributing any part of it in any form without permission. You are supporting writers and allowing Penguin to continue to publish books for every reader.

PENGUIN is a registered trademark and PENGUIN WORKSHOP is a trademark of Penguin Books Ltd. WHO HQ & Design is a registered trademark of Penguin Random House LLC.

Visit us online at penguinrandomhouse.com.

Library of Congress Control Number: 2023020553

Printed in the United States of America

ISBN 9780593520833 (paperback)   10 9 8 7 6 5 4 3 WOR
ISBN 9780593520840 (library binding)   10 9 8 7 6 5 4 3 2 1 WOR

The publisher does not have any control over and does not assume any responsibility for author or third-party websites or their content.

Contents

What Do We Know About the Chupacabra? . . 1

Cryptids 5

Making Headlines 13

Hairy Arms and Three Toes 29

Stealing the Life Force 48

Strange Discoveries 58

Bloodless Killers 68

The Trouble with Memory 75

Celebrity Chupacabra 83

Creatures from Outer Space 92

Timelines 106

Bibliography 108

What Do We Know About the Chupacabra?

On the morning of May 22, 1995, Don Francisco Ruiz walked outside his home in Humacao, on the eastern shore of Puerto Rico. Along with several beaches that draw a lot of tourists, Humacao boasts a large tropical forest and miles of flatland where people raise crops and animals.

Francisco was a rancher. When he went out to check on his animals that morning, his cattle were fine. But when he came to the enclosure where he kept his goats, he had a terrible shock. Three of his goats lay dead on the ground.

Francisco knelt down to examine them. Each one had what looked like puncture marks on its body. He thought an animal must have attacked them. But what kind of animal? He looked closer at the puncture wounds. Something had bitten his goats. But there was no blood on the ground or in the wounds. In fact, it looked as if the goats had been completely emptied of blood.

Don Francisco Ruiz knew a lot about animals. He was a farmer who knew what cattle ate and how goats behaved. But he likely did not fully understand what he was looking at.

What had happened that night? What kind of animal would do this to his goats? Francisco's own blood ran cold. He remembered reports in local newspapers and on the radio about farm animals with strange wounds, animals that had been found completely drained of their blood.

Although no one had seen the creature that did this, its name was being whispered all over Puerto Rico. In Spanish they called it the Chupacabra, or "goat sucker." Had this mysterious beast visited Francisco's farm?

CHAPTER 1
Cryptids

Today, anyone can look up information in books or on the internet about the animals in our world. We know why some eat plants while others eat meat. We understand why some have fur and some have scales, as well as where they live and how they travel. Even if you have never seen a penguin in real life, it is easy enough to find pictures or videos of them.

Long ago, this wasn't the case. People had no idea what kinds of creatures might live out in the parts of the world they had never seen. They imagined things like the ichthyocentaur (say: ICK-thee-o-CEN-tor)—part human, part horse, part fish, and able to play musical instruments—and creatures such as mermaids. Many beings that we know are imaginary today were once thought to be real. Mapmakers even sometimes drew mythical beasts and animals on maps to show people where they might be found. To someone who had never seen an elephant, that massive creature might seem as fantastical as an imaginary unicorn.

Ichthyocentaur

Even today, we can't be sure we've documented every animal that exists in the entire world.

Cryptozoology (say: KRIP-toe-zoh-AHL-o-jee) is the study of mysterious creatures whose existence hasn't yet been proven. *Crypto* means hidden and *zoology* is the study of animals. The animals that cryptozoologists search for are called cryptids because if they *do* exist in the world, they remain hidden. We don't yet have proof that they exist. Bigfoot, the hairy giant some say stalks the woods of North America, is a cryptid. So is Nessie, the underwater monster said to live in Loch Ness, a lake in Scotland. And so is the Chupacabra.

Loch Ness Monster

Cryptozoology is not considered a true science, because science studies things that can be observed. But many people still believe that reports about beings like Bigfoot should be taken seriously. After all, some animals that once seemed fantastic turned out to be very real.

Bigfoot

The narwhal is a type of toothed whale that lives in the Atlantic and Arctic Oceans. Because of the large tusk growing out of its upper jaw, it's nicknamed "the unicorn of the sea." In the Middle Ages (a period of European history from the late 400s to the 1400s), narwhal tusks were sold as actual unicorn horns. Some people believed they had magical powers.

Narwhals are very real. Yet every month, almost fifteen thousand people search the internet to ask: "Are narwhals real?" So even the existence of real animals that we have studied and documented is sometimes questioned.

Narwhal

The Kraken

For centuries sailors have told stories of a terrible sea monster that lived beneath the waves and attacked their ships with its giant tentacles. Scandinavians called the creature the kraken. In Japan it was known as Akkorokamui. The ancient Greeks called it *teuthos*.

With so many different cultures believing in it, could the kraken be real? It turns out yes, it was—and it is. The giant squid, as it's called today, lives deep in the ocean. The largest found so far was forty-three feet long and it weighed a ton (that's two thousand pounds!). One species of giant squid have long, powerful tentacles covered in suckers that are filled with teeth.

This is one monster that lives up to its legend!

In order to know that an animal definitely exists, we need proof, like the body of a giant squid or a captured narwhal. Without an actual sighting of the animal, any small bit of evidence will do: a strand of hair, a footprint, a nest, or a piece of skin or scale. That is what cryptozoologists search for. In 1995 they began looking for a new cryptid in Puerto Rico: the Chupacabra.

CHAPTER 2
Making Headlines

The town of Orocovis is located in the Central Mountain Range of Puerto Rico. The mountains cross the island from east to west, dividing it into northern and southern coastal plains. Because of its central location, Orocovis is sometimes called the Heart of Puerto Rico. Many people in the region make their living by growing wheat and coffee, or raising livestock like goats, sheep, and cattle.

Puerto Rico

Puerto Rico, in the Caribbean Sea, is made up of 143 islands of various sizes, but people only live on three of them. In total, it covers about 3,500 square miles. Nearly a quarter of the land is made up of steep slopes, with lowlands near the coasts. As of 2020, there were 3.286 million people living in Puerto Rico.

In the fifteenth century the island was called

Borinquén by the Taino, one of the two major groups of people who lived there (the other was the Caribs). European conquistadors—or conquerors—arrived from Spain in search of gold and other riches. They renamed the land Puerto Rico, Spanish for "rich port." The Spanish enslaved or killed many of the island's people and brought enslaved people from Africa to work for them. Many people living in Puerto Rico today are descended from a combination of the Indigenous groups, Europeans, and Black African people. In 1917, Puerto Rico became part of the United States.

In March 1995, several residents of Orocovis and the nearby town of Morovis woke up to discover some of their livestock had been killed. Ranchers went to their fields and found goats, sheep, and cows lying dead with small puncture wounds in their necks. Even more chilling, the

animals seemed to have been drained of blood. No one had seen any evidence of a person or animal that could have killed them. What was attacking their valuable livestock? By the time Don Francisco Ruiz found his own goats dead outside his home in Humacao, he had already heard about the attacks in Orocovis. Eventually the mysterious creature was given a name: Chupacabra, or "goat sucker."

Giving the creature a name made it easier to talk about. At first, the stories were spread from person to person. People passed on rumors that they heard from a cousin about a neighbor whose sheep had been killed by a monster. But soon the stories caught the attention of local newspapers. The publishers of those papers knew they could sell a lot of copies with headlines about a monster stalking the countryside.

Who Named the Monster?

The word *Chupacabra* needs no explanation when you know how the creature eats, but where did the name come from originally? The credit goes to Puerto Rican musician, writer, and comedian Silverio Pérez.

Born in 1948 in Guaynabo, Puerto Rico, Silverio was hosting a radio show in San Juan, Puerto Rico's capital city, in 1995. The conversation turned to the mysterious animal attacks that were going on. He referred to the creature as a Chupacabra—the goat sucker—as a joke, and the name caught on.

Silverio Pérez

In 1995, the most popular newspaper in Puerto Rico was *El Vocero* (say: el vo-CER-o), which means *The Spokesman*. Today, the paper publishes many types of news, but back in 1995 it was known for its focus on dramatic stories, especially violent ones, which it advertised in big red headlines.

The Chupacabra quickly became the star of *El Vocero*. The paper published story after story about the mysterious creature, often written by the same few writers. Reporter Ruben Dario

Rodríguez alone was responsible for nearly half the Chupacabra articles the paper ran. The more stories Rodríguez wrote, the more exciting the details became. The reporters at *El Vocero* didn't spend much time trying to prove whether or not the stories were true. They just repeated what people told them.

For instance, in November 1995, the paper claimed the Chupacabra had killed a cat and a sheep before swallowing an entire lamb whole.

No one actually saw the lamb being swallowed by anything. That was just the conclusion the journalists came to when the lamb couldn't be found. A few days later the Chupacabra was reported to have killed five chickens before placing a strange mark on the arm of a five-year-old girl whose parents owned the chickens.

The experience was said to have turned the girl into a genius! There were no follow-up stories proving that the girl was now brilliant, and no one examined the mysterious sign on her arm. They just moved on to the next story.

Whether or not people believed any or all of these stories, they definitely popularized the mysterious creature and sold newspapers. But if the Chupacabra was a real animal, some people wondered why it only started being widely reported in 1995. People had been farming in Puerto Rico for centuries.

Why did it take until the end of the twentieth century for the farmers to give a name to a creature that preyed on their livestock?

Those who believe the Chupacabra exists have explanations for all this. One of the most popular theories involves the government of the United States.

According to that theory, the Chupacabra was created by scientists working for the US government. The scientists created an entirely new animal, using features of other creatures that already existed.

This idea is similar to the plot of the American movie *Species*, which was released in Puerto Rico on July 7, 1995, when stories about the Chupacabra were already in the news.

The idea of US scientists conducting secret experiments in Latin America—those areas of the Americas where Spanish and Portuguese are spoken—is not new to many people who live there. That's especially the case in the Chupacabra's homeland of Puerto Rico, which is a US territory. The US Department of Agriculture's Forest Service has conducted many experiments in Puerto Rico's El Yunque National Forest, including tests related to atomic radiation. The government wanted to see what effect gamma radiation, a type of energy produced when an atom bomb is exploded, had on the plants there. Although many native plants survived, there is much scientists still don't know

about how these dangerous rays might have changed the forest, and the land surrounding it. The people of Puerto Rico were understandably angry.

El Yunque National Forest

The US government also stored and tested dangerous chemicals in Puerto Rico, including poisons that could seep into the ground and water and make animals and people sick. Many of these experiments were run out of a US navy installation in Vieques, which made local people unhappy about the installation's existence and distrustful of anything that might be going on there.

With this history, it's easy to see why Puerto Ricans could so easily believe that the mysterious creature killing their livestock was yet another example of the United States government using their home as a laboratory even when it put people in danger.

CHAPTER 3
Hairy Arms and Three Toes

People in Puerto Rico were beginning to understand what the Chupacabra did. But what, exactly, did it look like?

That depended on which report a person read. At first, people only saw the damage the Chupacabra left behind. The creature was long gone by the time its victims were discovered. The first published sighting of the creature itself came from the city of Caguas in November 1995. The witness claimed the monster had broken into his house and that it had hairy arms and huge red eyes. He said the Chupacabra ripped up a teddy bear before escaping through a window, leaving behind a puddle of slime and a piece of white meat.

Not long after that report, a resident of Canóvanas claimed he, too, had spotted the Chupacabra. He thought it looked like it belonged to the monkey family, but it wasn't a monkey.

It was about four feet tall and didn't have a tail. In both these descriptions, the Chupacabra walked upright on two legs.

One witness claimed the creature had a shaved head. Another described it as running "like a gazelle." A housewife claimed she'd looked the monster in the eye and said, "If you're the Chupacabra, you're a sorry excuse for a creature."

Another witness said the creature was covered in dense black feathers. Yet another said the skin on its body changed color from purple to brown to yellow, while its face stayed dark gray.

With such a variety of descriptions, it was hard to know what the Chupacabra really looked like. But over time, people started to agree on at least a few qualities of the creature, and a clearer image began to take shape. The Chupacabra was between three and five feet tall. It had large, slanted black eyes that sometimes glowed red, and it had pointed ears.

But one description appeared that became the most popular. Of all the people who claimed to have seen the Chupacabra, none was more influential than Madelyne Tolentino. Madelyne lived in Canóvanas, a town to the east of San Juan. She saw the creature in August 1995 when looking out her kitchen window. Madelyne gave the most detailed description of the Chupacabra yet.

According to her, the Chupacabra had dark eyes that spread around the sides of its head. It walked on two legs and was about four feet tall. It had thin arms and legs, with three fingers on each hand and three toes on each foot. It had no ears and no nose, but it did have two small airholes in its face. She also said that it had feathers or some kind of spikes growing out of its back.

Artist Jorge Martin sketched a picture of the Chupacabra based on Madelyne's description. The picture was reprinted in newspapers along with her story. Since no Chupacabra had ever been captured or photographed, that picture and Madelyne's description became the closest thing to a real Chupacabra that people had ever seen.

Jorge Martin

When people talked about the Chupacabra by the autumn of 1995, they would usually imagine the one Madelyne described.

By the start of 1996, sightings of the Chupacabra had become widespread, from its original home in the Central Mountain Range to the east coast. But this was still strictly a Puerto Rican phenomenon. Stories were still spread via word of mouth or the local news.

Central Mountain Range

But later in 1996, the Chupacabra caught the attention of Cristina Saralegui, an American journalist who had been born in Cuba. Cristina hosted a popular afternoon talk show filmed in Miami, Florida, called *El Show de Cristina* (*The Cristina Show*).

El Show de Cristina was broadcast on Univision, an American network with headquarters in New York City, and was watched in Spanish-speaking countries all over the world. *El Show* had an estimated one hundred million viewers worldwide!

Cristina Saralegui (1948–)

Cristina Maria Saralegui de Ávila was born in Miramar, Havana, Cuba. She and her family moved to Miami, Florida, in 1960. She attended the University of Miami, then started a career working as a journalist for magazines, including the Spanish-language version of *Cosmopolitan*.

In 1989 she began working in television, hosting her own talk show in Spanish, called *El Show de Cristina*. At the end of each show, she would give the audience a double thumbs-up and tell them to keep going forward and never step back. She hosted her show for twenty-one years, until November 1, 2010. In 2012 Cristina began hosting a radio show, *Cristina Opina*.

On one episode of the show, Cristina was joined by José "Chemo" Soto, the mayor of Canóvanas. The mayor had been leading weekly hunts for the Chupacabra, hoping to lure it out with live goats. But, he hadn't had any luck. He warned people that while the creature had only attacked animals so far, it might start hunting people any day.

José "Chemo" Soto

This single episode of Cristina's show made the Chupacabra an international sensation. Sightings of the animal were immediately reported in the United States and Mexico. And it didn't stop there. Soon there were sightings in other countries, too.

The first report of a Chupacabra appearance

on the mainland of the United States came from Miami, Florida, in March 1996. The beast itself wasn't seen, but a researcher named Virgilio Sanchez-Ocejo had made plaster casts of footprints left behind in the dirt. To him, the tracks looked like they belonged to an unknown creature, possibly not from this planet.

And the reported attacks and sightings of the Chupacabra in the United States continued, especially in areas where people spoke Spanish. A boy in Tucson, Arizona, claimed the Chupacabra broke into his house. He said the monster came right in the front door, slammed it behind itself, walked through the kitchen, and even sat on his bed before jumping out the window. The boy said the Chupacabra was about three feet tall with long arms and a red nose like a bird's beak.

In April, Patricia and Mario Mendez-Acosta, scientific investigators from Mexico City, began a search for the creature that was detailed in *Skeptical Inquirer* magazine. The pair kept watch in farmyards where it was said the Chupacabra had attacked animals.

They did catch some animals—but each time the attacker turned out to be some kind of dog.

A lack of proof didn't stop the Chupacabra's territory from expanding to include much of Latin America. Nine pigs were found dead in Brazil. Three hundred animals were said to have been killed in Chile. These were all countries where people spoke Spanish or Portuguese, languages close enough that speakers of one can often understand the other.

The new sightings also created more theories about how the creature could have so suddenly appeared. One popular story that spread all over Chile claimed that three Chilean soldiers had discovered not just one Chupacabra, but three. They seemed to be a family of a male, a female, and a cub. They were living near a mine north of the city of Calama.

The soldiers were not able to show their Chupacabras to the public, however, because soon after the discovery a team of scientists arrived from the United States. They worked for NASA, the National Aeronautics and Space Administration. Scientists at NASA study the vast unknown areas of space and planetary science.

The Chilean soldiers claimed that scientists took the Chupacabras away in a black helicopter. They were taking them back to the secret NASA lab where they had been born. Although

NASA is part of the US government, this lab was in the Atacama Desert in northern Chile.

It was said that the Chupacabras were part of an experiment for the purpose of space travel. Some people believed that the NASA scientists were trying to make an animal that could survive on the planet Mars!

Other rumors about lab-created Chupacabras gave the scientists a scarier mission: that their goal was to create some kind of an animal weapon that could fight for them in battle. But, just like in the story that had originated in Chile, it was believed that some of their animals escaped and were now living in the wild in Latin America.

What was the public's reaction to these reports? People who depended on farm animals to earn their living took the reports very seriously. Like Mayor Soto, they sometimes formed patrols to go hunting for the killers. But others thought the reports were just silly made-up stories. Some people didn't believe everything they read or heard about the Chupacabra, but they wondered if something strange was happening.

For many people in Latin America, however, the idea of a Chupacabra was scary but also a bit funny. Puerto Rico was proud that its local monster had become a sensation. By the late 1990s, there hadn't been any sightings of the Chupacabra in El Salvador. But a visiting tourist could find T-shirts for sale that showed the monster arriving in that country, often carrying a suitcase covered in stickers showing the places where he had already been seen, like Mexico, Nicaragua, and Argentina. Local citizens weren't sure *if* the Chupacabra would ever enter El Salvador, but they were ready to have some fun with his story in the meantime.

CHAPTER 4
Stealing the Life Force

Whatever people imagined the Chupacabra might look like, they knew what it liked to eat: blood. The Chupacabra was a bloodsucker, the characteristic that is often associated with vampires.

The word *vampire* did not exist in the English language until 1732, but bloodsucking creatures already existed in different myths from around the world. The ancient Egyptian goddess Sekhmet, who often took the form of a cat, was believed to drink blood

Sekhmet

after fighting in battle. The ancient Greeks told legends of the Lamia, who feasted on the blood of children as they slept. And the Estries of Jewish folklore needed to drink blood to survive.

Vampires can have many different abilities depending on who is telling their story. Some can shapeshift into different animals, some can't be seen in mirrors, some can't be outside during the day. But there are two things that define something as a vampire: It drains the "life force" of its victims, and it is blamed for unexpected deaths.

The most famous vampire in the world was created by Bram Stoker in 1897. His name was Count Dracula.

The Chupacabra was far from the first bloodsucking creature to exist in Spanish-speaking countries. And mysterious monsters already held a special place in Puerto Rico. In the annual celebration of Carnival, people dance and sing in the streets. Some dress as mythical characters known as *vejigantes* (say: veh-hee-GAN-tays). They wear elaborate papier-mâché masks with horns, fangs, and bulging eyes. As fierce as they are, the masks are also beautiful.

Some families pass the art of mask-making down through generations. The masks have become so tied to Puerto Rican culture that tourists sometimes buy them as souvenirs to display back home.

Bram Stoker (1847–1912)

Abraham "Bram" Stoker was born in Dublin, Ireland. He was the third of seven children, and graduated from Trinity College in Dublin. He loved the theater, and eventually moved to London to work at the Lyceum Theatre there. Bram spent years studying the vampire folklore of Central and Eastern Europe to create the polite but deadly fictional character Count Dracula, who still influences modern stories about vampires. Dracula's story has been adapted many times for the stage, film, and television.

Puerto Rico and some Latin American countries have a colonial past. Colonialism is when a powerful country takes control of a smaller country and its natural resources, like gold or wood, to enrich itself, leaving little behind for the people of the colony. In fact, sometimes the Indigenous people themselves are the natural resource for the more powerful countries who enslave them.

The behavior of the colonizing country has a lot in common with a vampire. It sucks the life force out of the smaller country. Perhaps that's why Latin American countries have so many vampire myths.

The Chupacabra wasn't even the first legendary bloodsucking creature to prey on farm animals in Puerto Rico. In 1975, farm animals in the town of Moca were said to have died in attacks similar to those by the Chupacabra twenty years later. Some people claimed to have heard loud

screeches or the flapping of wings when the attacks were taking place. The monster became known as the Vampire of Moca, but could it have been the same beast we now call the Chupacabra? Could bloodsucking cryptids really exist?

Vampires are supernatural creatures that feed on blood. They only appear in movies and stories. But there are animals in the real world that actually do feed on blood. Some of the most common are insects, such as fleas and mosquitoes, and worms, like leeches. And some are mammals, like the vampire bat. Vampire bats have hair or fur rather than feathers or scales, and they are born live rather than being hatched from an egg. Human beings are also mammals. But vampire bats are the only mammal to live entirely on blood.

A vampire bat must eat every two days. It will feed off everything from cattle to reptiles to birds and even, in rare cases, people. A bat usually lands on an animal that is sleeping and bites it with its tiny, sharp teeth. The vampire bat then licks the blood from the bite with its tongue, which is specially shaped to channel blood fast. Usually, a tiny cut like the one made by a vampire bat will stop bleeding quickly and form a scab. But the saliva (spit) of vampire bats contains a chemical that will keep the cut from healing before the bat can finish eating. It also contains a chemical that keeps the cut from being painful to the animal, so often the bat's bite will not even wake up its victim.

Vampire bat

Vampire bats have other special abilities to help them find food. They can actually sense a sleeping animal breathing from the air while they are flying. A single bat might return to the same animal to feed several nights in a row. Once the bat has landed on an animal, it uses heat sensors located around its nose to find a place on the animal's body where blood is close to the skin.

When it is finished feeding, the vampire bat uses its strong hind legs and its thumb to take off quickly and fly back into the air. It can walk gracefully on its legs and wings. That comes in handy when the bat wants to sneak up on an animal from the ground.

Almost everything about the vampire bat helps it to find, eat, and digest the blood it needs to live. If the Chupacabra also lives on blood, it would need to be able to hunt and feed like the tiny vampire bat. Researchers would need to examine the body of a Chupacabra to see if that were true.

In August 2000, a farmer in Nicaragua claimed to finally have the body of a Chupacabra to study. Would it hold the answers that scientists needed?

CHAPTER 5
Strange Discoveries

Five years after the Chupacabra began hunting animals in Puerto Rico, a farmer named Jorge Talavera, of Malpaisillo, Nicaragua, claimed to have killed one. The Chupacabra had been attacking farms in town. Jorge said it could kill an average of five sheep and a goat every night, and he was determined to track it down. On August 25, 2000, Jorge and a friend stayed up to keep watch over his sheep. Late that night, they heard one of his goats cry out. Jorge saw strange creatures moving among the sheep and fired his gun at them. He hit one of them, but it was still able to run away.

Three days later, a ranch hand on Jorge's farm found the animal's remains in a nearby cave.

Whatever it was, it looked like it had been dead for a few days. Other animals had already eaten some of it, but there was still plenty left to examine. The ranch hand called Jorge to come see it. When he looked at it, Jorge agreed that this certainly was the creature he had shot.

It had no tiny ears, no body hair, and didn't look like any animal they had ever seen. It didn't look like the popular descriptions of the Chupacabra, either. Nevertheless, Jorge was certain that's exactly what it was.

Jorge's story made international news. People all over the world offered opinions on what he had found. Some said it was a strange animal that had escaped from a circus. Others thought it was an unknown species, perhaps from somewhere in Africa. A local veterinarian thought it might be a combination of many species, maybe created by scientists in a lab.

The remains were sent to the National Autonomous University of Nicaragua and examined by a team of scientists. They finally declared that the animal was . . .

. . . a dog.

An ordinary dog. It probably suffered from mange, a skin disease that makes animals lose their hair. As for the animal's mouth, there was nothing about it that would give it the ability to suck blood—even through a straw.

A dog with mange

Jorge Talavera might have been the first person to say he found a dead Chupacabra, but he would not be the last. In 2002, a man found the remains of a strange-looking creature lying on a mesa outside Albuquerque, New Mexico. Its face looked human, but it had a pointy head, stubby wings, and a tail. He thought it might be a Chupacabra. He gave the remains to a friend, who eventually gave it to New Mexico's Department of Game and Fish. He hoped they could tell him what it was.

Dried skate known as a "devil fish"

The carcass was obviously no dog. But it wasn't a Chupacabra, either. It was a skate! Skates are fish. They are related to stingrays. Skates are also called "devil fish" because of how they look—and how they could be made to look with a little help. Fishermen sometimes trim off the "wings," the only part of the fish that is considered edible, which alters their bodies to look like devils, angels, or dragons. It was no wonder this body in the desert looked like a monster. It had been altered to look like one.

Over the next decade, several other

Chupacabra-like bodies were found. But every time, they turned out to be some kind of dog, or coyote, or a mix of both.

Not everyone was convinced by the researchers' conclusions. Jorge Talavera accused the scientists at the Nicaragua university of stealing the Chupacabra he had shot and secretly switching it with a dog. He wasn't sure exactly why they would do such a thing, but he felt certain that the animal he shot and his ranch hand found days later was not a dog. After all, it had sucked the blood out of his sheep. He had seen that himself.

Or had he?

Coyotes

The coyote is a relative of the type of wolf native to North America. It has also been known as a prairie wolf and a brush wolf. Coyotes mostly eat wild animals like reptiles, rabbits, and deer, but they can hunt domestic animals like chickens, sheep, and goats as well.

Coyotes, dogs, and wolves are so closely related that they can parent cubs together. The cubs they create are called hybrids because they are a mixture of more than one animal. Coyote-wolf hybrids are called coywolves, and coyote-dog hybrids are coydogs. Coyotes are known for the many sounds they make, especially their howl, which—unlike the howl of a wolf—is punctuated by yips and barks.

CHAPTER 6
Bloodless Killers

Jorge Talavera knew his sheep had been drained of blood the minute he saw them. So did other farmers whose animals were attacked. Newspaper reports, especially those in *El Vocero*, took the farmers at their word. They often added dramatic language, proclaiming that "not a drop of blood" had been left behind inside the animal victims.

But in fact, none of the animals killed were ever examined to see how much blood they had lost. Why would people think, just by looking at the animal, that it had no blood left inside it?

There are a few reasons a person might jump to this conclusion. First, there is what the farmer expects to see. He knows his animals have been killed by some predator, and he expects that an attack would leave a lot of blood on the ground. After all, if an animal gets cut or bitten, it bleeds just like a person would. When there are no pools of blood on the ground and no blood even on the animal, it may seem as if the blood has somehow mysteriously disappeared.

When an animal is alive, its heart pumps blood throughout its body. The blood circulates from the heart through the arms, legs, and head. That's why it rushes out when the skin is cut. But when an animal dies, its heart stops beating. The blood stops circulating through the body. It sinks like any liquid would, and usually ends up pooling in whatever part of the animal is closest to the ground.

Even if a farmer did cut the animal's skin looking for blood, he probably would not find any. It's surprisingly difficult to know how much blood has been left in the body of a dead animal—too difficult for a person to be sure just by looking at it. Of course, once the Chupacabra's habits became known, people were even more quick to assume that their animals had been drained of blood.

There is another thing farmers often found on Chupacabra victims: Many of the animals had puncture wounds on their neck, just like the marks vampire fangs make on the necks of their victims in movies. These two little holes seemed far too small to kill an animal as big as a goat, sheep, or cow on their own. Besides, if the animals were attacked by some other predator, one that didn't feed on blood, why hadn't the animals been eaten?

Once again, these witnesses might be influenced by what they *expect* to see. They think an animal like a coyote would only kill a goat in order to eat it. But in fact, the attack of a coyote or some other doglike animal would look exactly like an attack of a Chupacabra. Coyotes usually kill animals by biting the throat behind the jaw and below the ear. That bite to the neck stops the animal from breathing.

After biting an animal's neck, a coyote might just run on to the next nearby animal and bite it, too. A single coyote might kill a dozen animals before getting tired or deciding to eat one.

However mysterious and strange they seemed to some of the farmers, these kinds of animal attacks aren't uncommon in nature. Perhaps the witnesses just didn't understand what they were seeing. Even when a person is trying to be accurate, they are often mistaken about what they saw. Or *when* they saw it. Or *if* they even saw it at all.

CHAPTER 7
The Trouble with Memory

By this time, many different types of animal bodies had been found that people assumed were Chupacabras, but when people imagined the creature, they still usually thought of the sketch based on Madelyne Tolentino's description in 1995—the one with elongated dark eyes and long, skinny fingers. Perhaps it was even more memorable because moviegoers at the time were watching a similar creature in theaters.

Madelyne Tolentino

Species was a monster movie about a creature

called Sil, a half-human, half-alien being created in a lab. Sil looked a lot like what Madelyne had described. She had the same eyes, the same long fingers, the same spikes coming out of her back. Neither had ears. When Sil killed something, she did it by sucking out its blood and organs.

Sil even shared other qualities with Madelyne's description of the Chupacabra that weren't widely known. For instance, Madelyne told some Puerto

Rican researchers that the Chupacabra had left behind some kind of slime on the animals it killed. She said some of the slime was sent to a lady in Pennsylvania who had it analyzed, and that analysis showed the slime seemed to come from somewhere other than Earth.

No records exist of the slime being analyzed. The lady Madelyne claimed analyzed it denied even having heard of it. Sil, the fictional monster of *Species*, did leave behind slime. Sil behaved like Madelyne's Chupacabra in other ways as well. She hissed and could jump great distances.

Species came out in theaters in Puerto Rico a few weeks before Madelyne claimed to have seen the Chupacabra in her yard. In an interview, she even talked about seeing the movie. She thought it did a good job making Sil look like the Chupacabra. She thought the plot was based on current events. The first scene even took place at an observatory in Puerto Rico.

But the monster Sil was not based on the Chupacabra. She was created by an artist named H. R. Giger before any reports of the Chupacabra appeared.

Does this mean Madelyne was simply influenced by the movie *Species* and that she was lying when she described the Chupacabra she saw? Maybe not.

It's not unusual for a person to remember—or think they remember—something that never happened, or something they only heard about, or saw in a movie.

People are able to create "false memories" even though they believe what they are saying is true. The human brain is capable of filling in gaps in memory with false information, including images, sounds, and emotions, that may feel very real.

H. R. Giger (1940–2014)

Hans Ruedi Giger was born in Chur, Switzerland. His father wanted him to be a pharmacist, but instead he studied architecture and industrial design at the School of Applied Arts in Zurich, Switzerland. There Hans became known for his weird drawings, which blended human beings with machines. His artwork appeared on posters and magazines. In the 1970s, when people bought music on vinyl records and tape cassettes, Hans's artwork often appeared on album covers. His style eventually brought him to Hollywood to work on movie special effects. His most famous creation was the terrifying monster in the movie *Alien*, which won him and his team an Academy Award for Visual Effects in 1980.

This could explain why Madelyne's memory of seeing the Chupacabra so closely resembled the movie she saw. It would also explain how her description of the creature was so detailed, as if she had time to study it up close. Madelyne could "remember" that whatever she saw through the kitchen window had elongated eyes and skinny fingers. She could "remember" collecting slime to be analyzed. She remembered these things, even if they never happened.

CHAPTER 8
Celebrity Chupacabra

The Chupacabra seemed to be everywhere in Latin America in the mid-1990s. Today, it's far more rarely reported. But just because it is no longer making headlines doesn't mean the creature has been forgotten. It became an international star very quickly, and it's still one of the most well-known cryptids in the world. Its name is almost as recognizable as Bigfoot and the Loch Ness Monster.

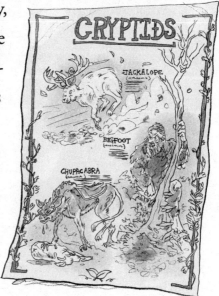

It's still easy to find the Chupacabra on T-shirts and other clothing. There are Chupacabra card games, models, and toys, including cuddly stuffed animals. It's even featured in picture books for children written in both English and Spanish.

Rudolfo Anaya

New Mexican author Rudolfo Anaya wrote his adventure story *Curse of the ChupaCabra* and its sequel, *ChupaCabra and the Roswell UFO*, for young adults. The Fantastic Four, a team of comic book superheroes, have faced off against the monster in Puerto Rico.

The Chupacabra has even been the subject of popular music. In 2021 the Randy Rogers Band joined with Mexican band La Maquinaria

Norteña to release a song called "Chupacabra." It had verses in English and Spanish, including, "You hit the floor like it's a piñata, roll your hips like an enchilada, you shake it out like it don't mean nada... that's how you do the Chupacabra!"

Clearly, the Chupacabra doesn't always have to be frightening. In fact, Spanish-language movies that feature the monster are often comedies. But the Chupacabra has turned up in scary stories as well. Just a couple of years after it was named, the Chupacabra was the subject of an episode of a very popular American TV show, *The X-Files*.

In the episode, called "El Mundo Gira" ("The World Turns"), a migrant worker from Mexico living in California is killed. It looks like something might have tried to eat her as well, causing some people to blame the Chupacabra for her death. But the circumstances of her death don't quite match the real-life reports of how the creature kills. Those details had already become so well-known that the show's writers expected many people to be familiar with them.

In the movie *Chupacabra Terror*, released in 2005, a cryptozoologist manages to capture the creature on a Caribbean island. He tries to bring it back to the United States by sneaking it onto a luxury cruise ship. Naturally the monster gets loose and puts all the passengers in danger.

The X-Files

The X-Files premiered on September 10, 1993. It focused on two Federal Bureau of Investigation agents, Fox Mulder and Dana Scully, who investigated cases that seemed to have supernatural or alien origins. Mulder believed that alien beings and other monsters already existed on Earth, and the government was hiding their presence from the public. Scully, a doctor, looked for more scientific explanations for the cases they encountered. At the height of its popularity, *The X-Files* had an audience of over nineteen million viewers each week in the United States. The show ended in May 2002, but it inspired two feature films, *The X-Files* comic books, and a revival season in 2016.

Fans will still recognize the series' tagline: The Truth Is Out There.

In *A Mexican Werewolf in Texas*, from that same year, a group of American teenagers track down the Chupacabra that's attacking their town.

Long after the Chupacabra has seemingly stopped preying on animals in Latin America, it remains a household name all over the world. There was even a Chupacabra repellent spray available on the internet! Its creator swore that he'd been using it for decades and had yet to meet the monster, and so he assumed it worked very well.

But while some people may be wondering how to keep the Chupacabra away here on Earth, others are looking to the skies.

CHAPTER 9
Creatures from Outer Space

The story of the second-most-popular explanation for where the Chupacabra came from begins far from the United States, in outer space.

Even before the Chupacabra became famous, Puerto Rico had a special relationship with UFOs, or Unidentified Flying Objects. That is the name given to strange objects seen in the sky that cannot be explained. (They are sometimes now called Unidentified Aerial Phenomena.) UFO enthusiast Carlos Torres claims that hundreds of UFO sightings have been reported in Puerto Rico since

the 1930s alone. Some of these sightings have even been caught on video, though of course no one knows what they really are.

UFO sighting

Many believe that in the deep waters just off the coast of the island, there is a USO, or Unidentified Submerged Object, that could be a spaceship. USOs are mysterious objects seen underwater that cannot be identified or explained. The area is located in the region of Lajas, not far from the Laguna Cartagena National Wildlife Refuge. There are those who say that one or more sunken ships have been spotted flying in and out of the water, which has led some local residents to believe that there could be a base beneath the waves where ships can take off and land. People have also reported ships flying in and out of Lake Cartagena, not far from the wildlife refuge.

About thirty-five miles away from Lajas, the US National Science Foundation is also looking to the sky and studying outer space at the Arecibo Observatory.

On November 18, 1995, when there were many reports of Chupacabra attacks in the local news, an unidentified disk was seen hovering over the antennae of a radio station in central Puerto

Rico. The station's radio controls started behaving strangely for no reason. Even more shocking, an old piece of equipment from 1957 that was stored at the station was said to have turned itself on without even being plugged in.

Many believe reports like these are evidence of alien visitors, and that those visitors are responsible for the Chupacabra. Perhaps, just like

in the movie *Species*, the Chupacabra was part alien, created by Earth scientists with help from extraterrestrial ones. Others have suggested that the Chupacabra is itself some kind of alien probe, sent to collect blood from Earth creatures so scientists from other planets could study it. Others thought the Chupacabra had been brought to Earth to test the atmosphere so alien beings could then visit the planet safely. Or maybe the creature was an extraterrestrial pet who escaped while its alien owners were visiting Earth.

Today, reports of Chupacabra attacks are rare. The last one may have been in 2010, when two men accused the cryptid of killing at least twenty chickens in Horizon City, Texas.

The Arecibo Telescope

The Arecibo Observatory was home to a thousand-foot-wide telescope. When it was built in 1963, it was the largest telescope in the world, with a reflecting dish covering 118 acres. The telescope was used to study astronomy, the atmosphere, planets, and meteors. Tropical areas like Puerto Rico are good spots for telescopes because the planets are highly visible from that part of the earth. One reason Arecibo was chosen over other tropical spots like Hawaii was that a natural sinkhole there (a large depression in the ground) could easily fit the telescope's dish. Hundreds of scientists from all over the world use the Arecibo Observatory for research.

One of the things the Arecibo telescope was used for is to search for extraterrestrial or "outside of Earth" intelligence. In 1974, the observatory sent a radio message about life on Earth to a distant star cluster. The Arecibo is the very observatory used in the opening scenes of the movie *Species*. On December 1, 2020, the platform supporting the giant telescope dish collapsed and destroyed it. The dish could not be repaired, but the observatory itself reopened in March 2022. Many tourists eagerly signed up to visit, and scientists continue to do research at the site.

But people still report seeing creatures they believe *might* be the Chupacabra. After all, no one yet knows what the Chupacabra really looks like. Perhaps one day someone will be able to capture one and study it. Or perhaps the

Chupacabra has eaten its fill and gone into hiding where cryptozoologists and government scientists can't find it. But we know they will certainly keep looking!

Conspiracy Theories

Many ideas about the origin of the Chupacabra rely on conspiracy theories, beliefs that something is the direct result of a secret plan. Although conspiracy theories often sound logical at first, they're generally based on emotions like anger or fear, not facts. Sometimes, when people don't understand something or simply don't like the way things are, they develop a theory (an opinion or explanation) that they say is the result of a secret plan carried out by powerful people.

Conspiracy theories come about because it can be easier to think that bad things in the world are part of a behind-the-scenes plot rather than to accept things as they are.

Timeline of the Chupacabra

- **1974** — The Arecibo Observatory sends a radio message to a distant star cluster
- **1975** — The creature known as the Vampire of Moca is said to kill animals in the town of Moca, Puerto Rico
- **1995** — An unknown animal kills farmer Francisco Ruiz's goats in Humacao, Puerto Rico
  - *El Vocero* publishes the first accounts of the Chupacabra
  - Madelyne Tolentino gives a detailed description of the Chupacabra
  - The Chupacabra is reported to enter a house in Caguas, Puerto Rico
- **1996** — The Chupacabra is a featured story on *El Show de Cristina*
  - A Chupacabra footprint is supposedly found in Miami, Florida
  - A boy in Tucson, Arizona, reports seeing the Chupacabra in his bedroom
- **2005** — The movie *Chupacabra Terror* is released
- **2013** — The Fantastic Four fight the Chupacabra in the comic book *Fantastic Four: Island of Death*
- **2020** — The platform supporting the telescope at the Arecibo Observatory collapses
- **2022** — The Arecibo Observatory visitors' center reopens

Timeline of the World

1974 — French acrobat Philippe Petit walks across a wire strung between the Twin Towers in New York City, 1,350 feet above the ground

1976 — The United States celebrates its bicentennial as the country turns two hundred years old

1983 — McDonald's adds Chicken McNuggets to its menu

1987 — The first Starbucks outside the United States opens in Vancouver, Canada

1990 — Tim Berners-Lee creates the first web server, the foundation for the World Wide Web

1994 — Angela Berners-Wilson becomes the first female priest of the Church of England

2003 — China launches the first manned space mission of the spacecraft *Shenzhou*

2006 — The International Astronomical Union downgrades the status of Pluto from planet to dwarf planet

2009 — Sonia Sotomayor, daughter of Puerto Rican parents, becomes the first Latina justice of the Supreme Court

2011 — NASA launches *Juno*, the first solar-powered spacecraft, to explore the outer planets of the solar system

2019 — The first case of COVID-19 is identified in Wuhan, China

Bibliography

***Books for young readers**

Akers Gozdecki, Kristen. "Puerto Rico UFOs." Clip from *UFO Files*, "Deep Sea UFOs" (season 3, episode 2). History. January 9, 2006. https://www.history.com/videos/puerto-rico-ufos.

Encyclopedia of the Unusual and Unexplained, Creatures of the Night: Chupacabra. http://www.unexplainedstuff.com/Mysterious-Creatures/Creatures-of-the-Night.html.

*Ha, Christine. *Chupacabra*. Mendota Heights, MN: North Star Editions, 2022.

Radford, Benjamin. *Tracking the Chupacabra: The Vampire Beast in Fact, Fiction, and Folklore*. Albuquerque: University of New Mexico Press, 2011.

Staff Report. "Report: Chupacabra Attacks Farm Animals." *El Paso Times*, June 28, 2016. https://www.elpasotimes.com/story/news/weird-news/2016/06/28/report-chupacabra-attacks-farm-animals/86456618/.

US Department of Agriculture Forest Service. "The US Military and El Yunque National Forest." https://www.fs.usda.gov/detail/elyunque/learning/history-culture/?cid=fseprd726155.